This item no longer
belongs to Davenport
Public Library

JONAH WINTER JEANETTE WINTER

Oil

Beach Lane Books

New York London Toronto Sydney New Delhi

From deep inside the earth it comes,

hot and black, black and hot,
up through the ground of the Far Far North,
pumped by machines, all day long, all night long,

day after day, year after year,
pumped by machines designed and run by humans,
all day long, all night long, oil

pumped into a giant pipeline
that crosses 800 miles of wilderness—oil

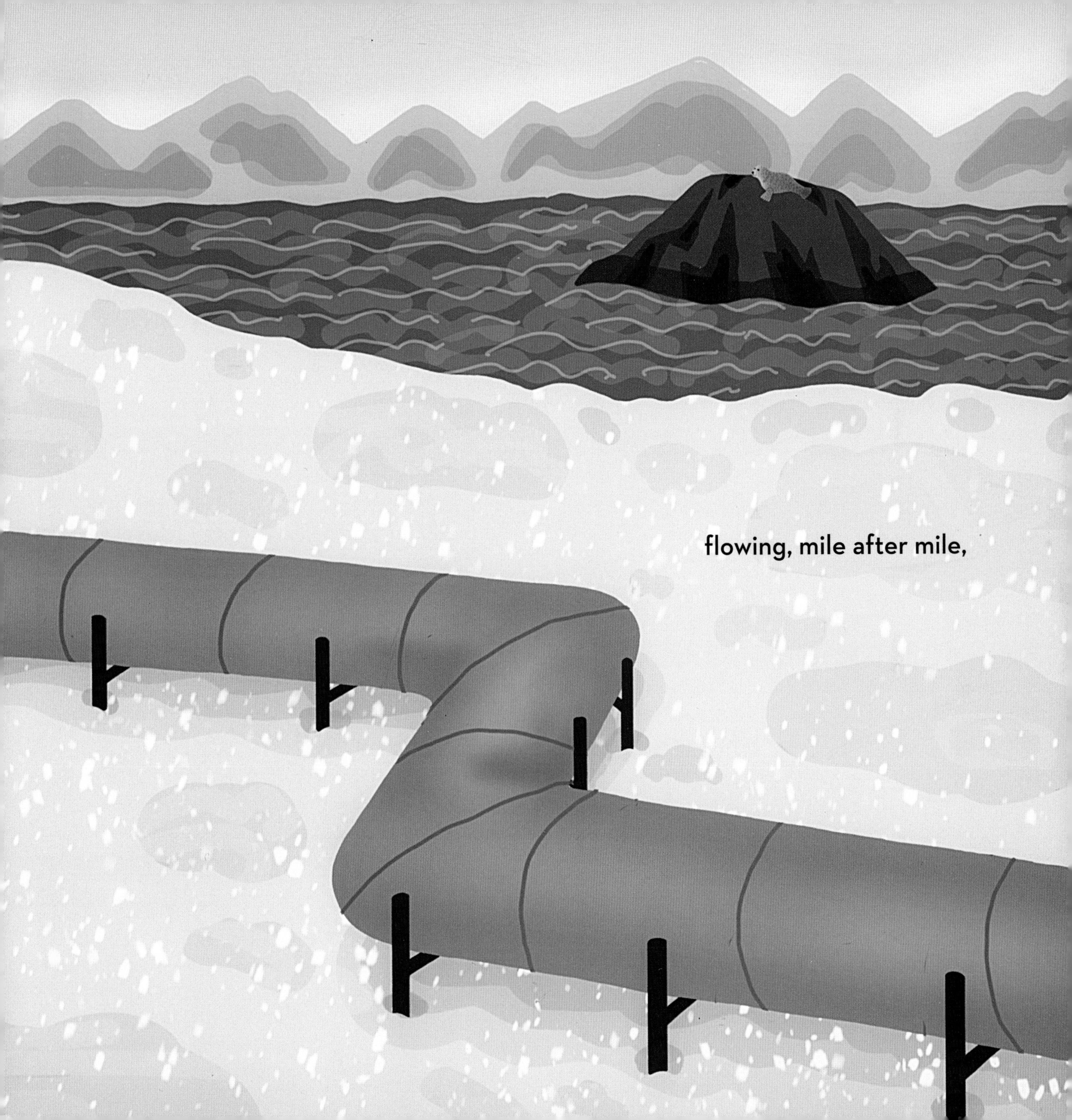

flowing, mile after mile,

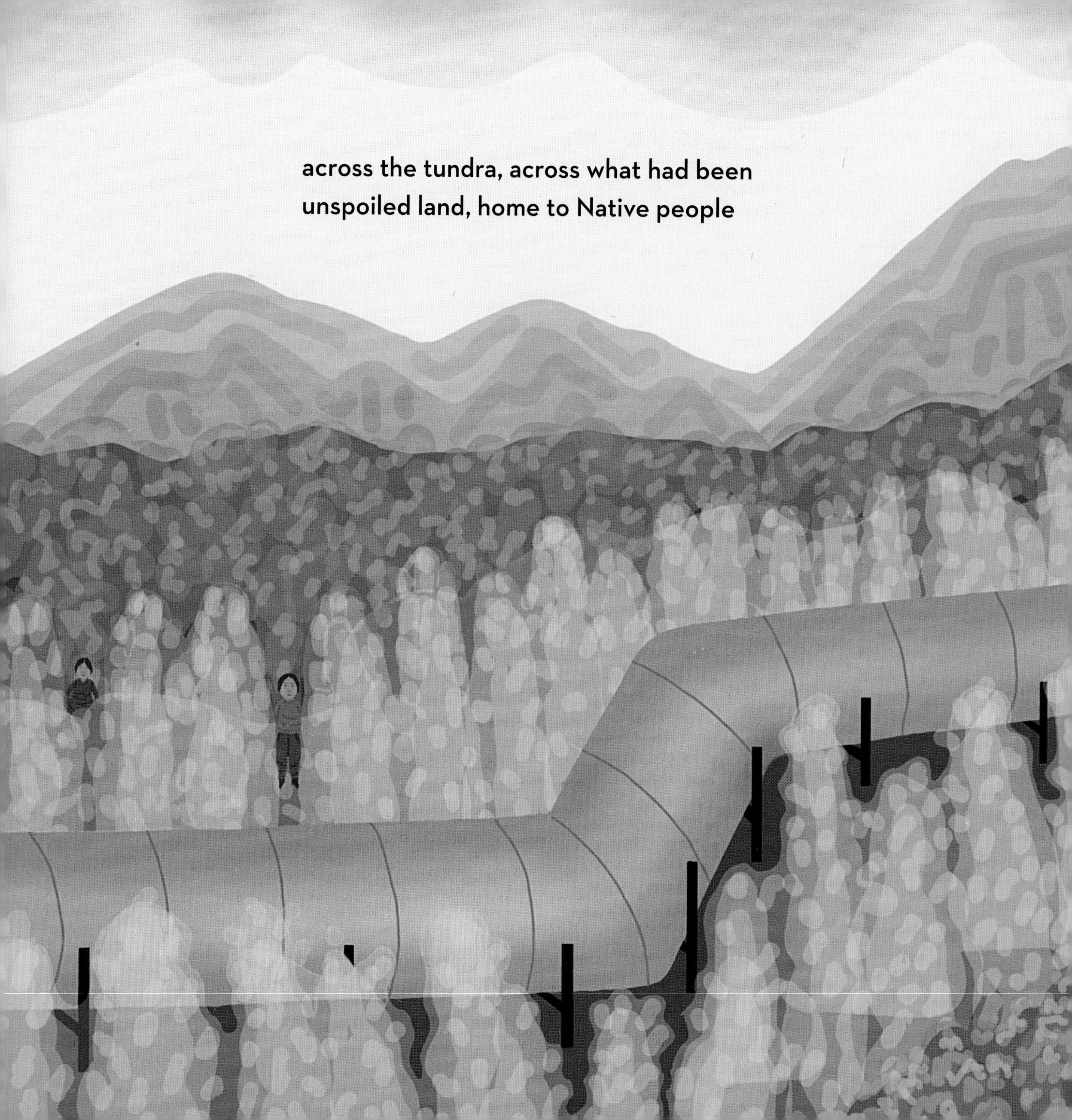

across the tundra, across what had been
unspoiled land, home to Native people

and thousands of caribou, oil
flowing onward

to a port on the ocean, oil

pumped onto enormous ships
to be transported farther still.

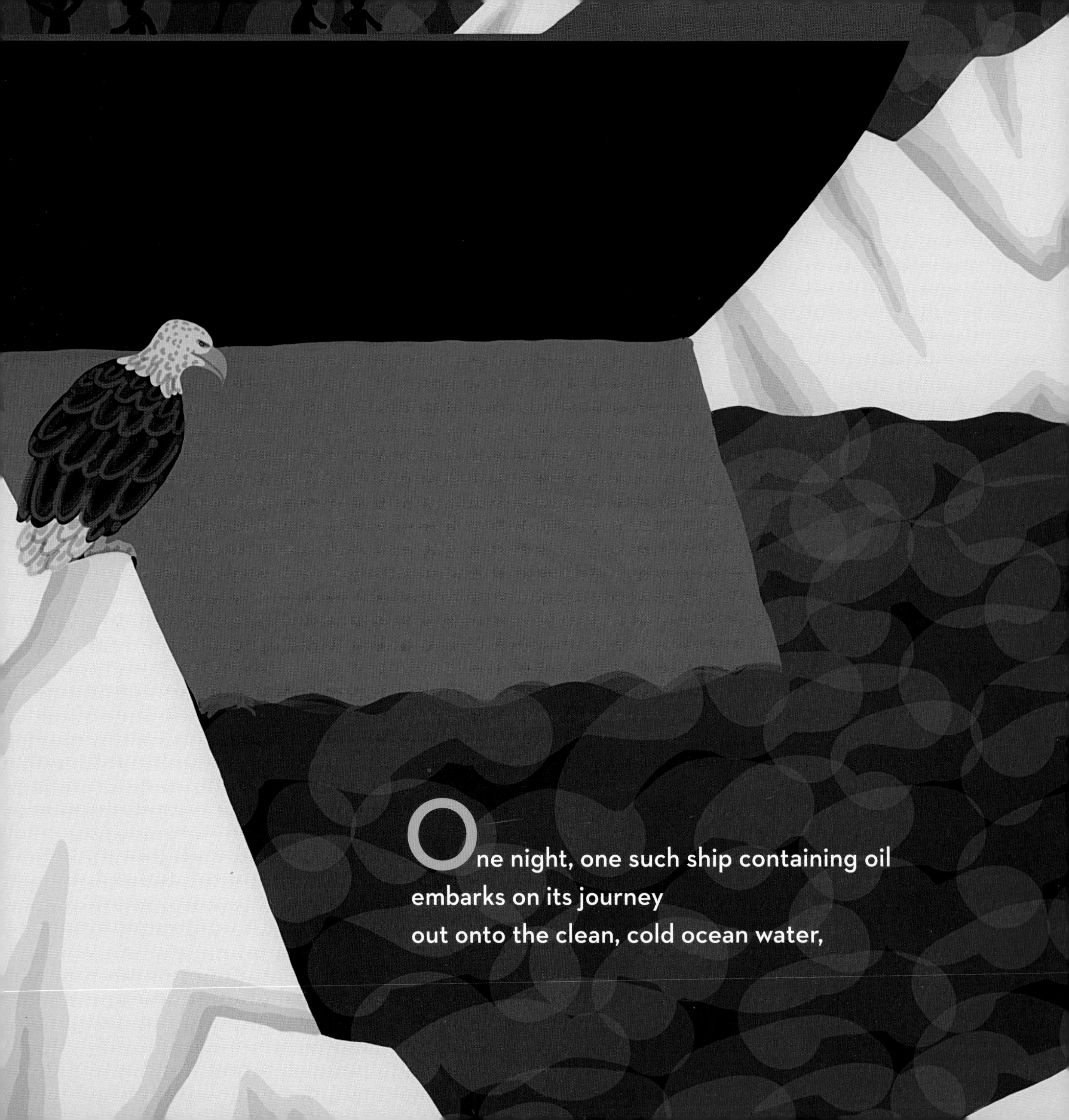

One night, one such ship containing oil
embarks on its journey
out onto the clean, cold ocean water,

gliding past gigantic icebergs,
moving swiftly through the night,
and then:

CLANG!

CRACK!

The ship runs aground—
onto a reef beneath the surface of the ocean.

And just like that, oil **GUSHES**
out of holes, some of them
as big as a house—oil

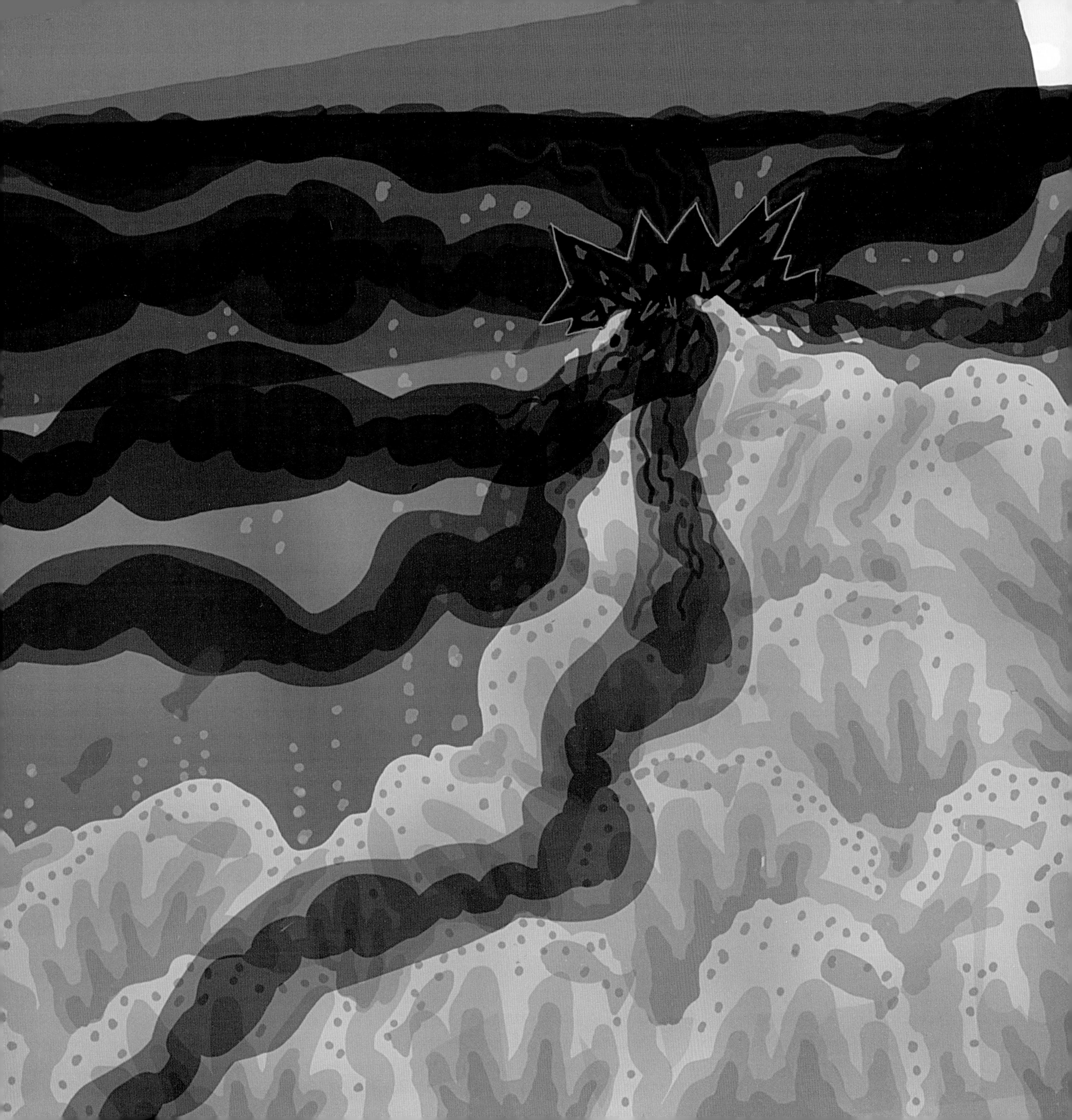

bubbling up to the
surface of the water,

THICK . . . BLACK . . . OIL . . .
spreading . . .

till black waves fall with a THUD onto rocky beaches.

Rocks that had been clean for millions of years
are now black and glistening with oil.

But some of the rocks aren’t rocks at all.

They are seabirds

and sea otters—

thousands of them, dead and dying.

There are some people who try to help.
They bring the struggling otters and birds to a rescue center.
They bathe the birds and otters, wash away the oil.

But most of the animals die.

And the oil

keeps

spreading

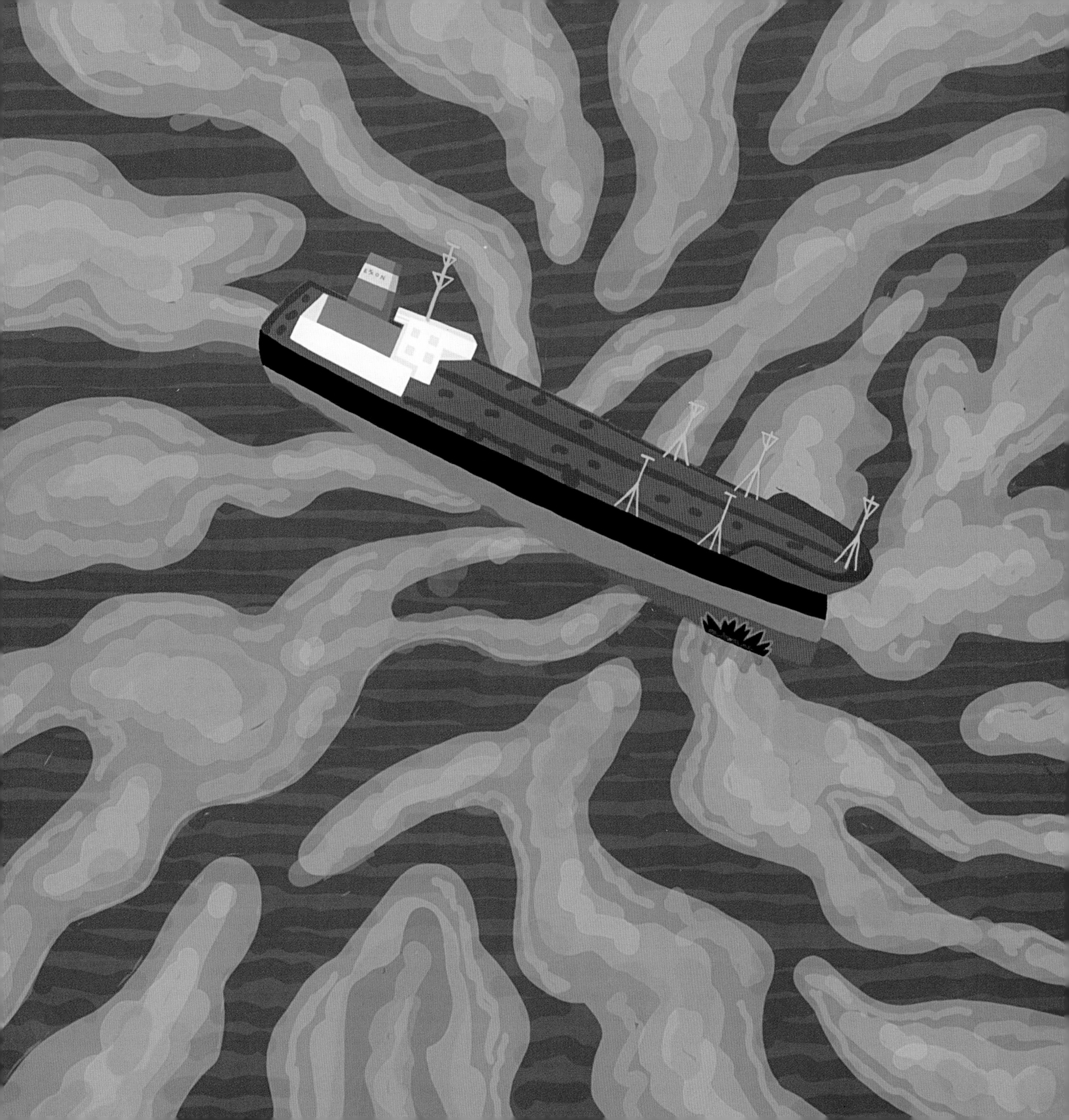

and spreading

and spreading, for days,

weeks,

m o n t h s —

till the oil covers thousands of miles of ocean.

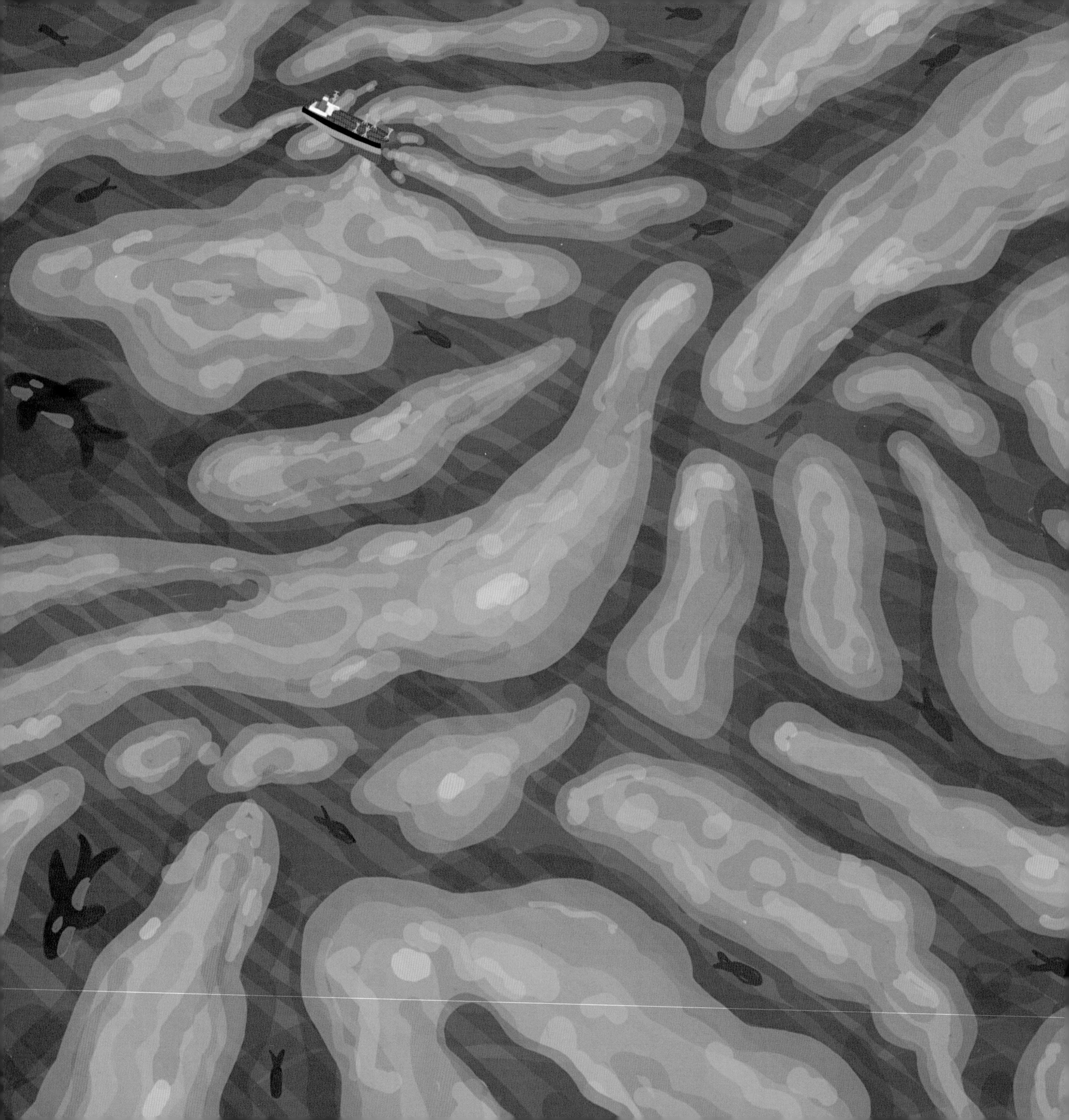

Thirty years later,
there are thousands of sea otters once again.
There are thousands of seabirds once again.
But this place is not the same.
The herring are gone.
Many of the killer whales are gone,
never to return.
For Native people,
who have fished here for thousands of years,
things are not the same—their way of life
still has not recovered.
And the beaches may look clean,
but if you lift a rock . . .

oil
seeps
up.

AUTHOR'S NOTE

This is a true story that took place in Alaska. In 1977 the Trans-Alaska Pipeline System was completed: an 800-mile pipeline originating on Prudhoe Bay and crossing over what had been pristine land occupied by Native Alaskans for centuries and home to many different animals. The pipeline was designed to transport the oil from Prudhoe Bay to the harbor of Valdez on Prince William Sound, where it could be transported on ships. On March 24, 1989, a ship full of oil called the *Exxon Valdez* ran aground in Prince William Sound on Bligh Reef at 12:04 a.m.—causing what at the time was the largest oil spill in US waters. Eleven million gallons of oil ultimately spilled from the ship into the ocean. The oil spill covered an area of 11,000 square miles—and washed up on 1,300 miles of shoreline.

There is no exact count on how many animals died as a result of the spill. However, 36,000 dead seabirds were recovered—though it is thought that somewhere between 350,000 and 390,000 seabirds died. Nearly 1,000 dead sea otters were recovered—most of the recovered living otters died soon after being rescued. There is no telling how many more thousands died though. All the eagles that had lived in this area were killed—and 151 of them were found. All in all, 90 different bird species suffered deaths due to the oil spill—including murres, kittiwakes, shearwaters, puffins, cormorants, loons, petrels, sandpipers, auklets, and peregrine falcons. One pod of killer whales living in Prince William Sound lost 15 of its 22 members, and there have been no calves born since the oil spill—meaning that this pod will someday disappear. Gray whales, sea lions, seals—all suffered injury and death from the oil. There is no estimate on how many fish died—probably in the millions. But the herring that had been plentiful in this area were completely wiped out.

Exxon, the oil company, was in charge of the cleanup. They only cleaned up approximately 14 percent of the oil—even though they spent 2.5 billion dollars on the cleanup and hired more than 10,000 workers to do the cleaning and spent several months on it. The problem was, and the problem still is, that you cannot effectively clean up a spill that big. The only way this cleanup could have been improved is if there had been an effective plan for cleanup in place or adequate equipment and know-how for dealing with such a catastrophe. Alas, there was no plan, very little equipment, and hardly any know-how. No one had much of a clue what to do—not Exxon, not the Coast Guard, not the Fish and Wildlife Service of the US government, not the people who ran the state of Alaska. They argued with one another. They wasted time. It was often not clear who was in charge. The animal-rescue people were ultimately blocked by both Exxon and the US government from rescuing any more animals. The more animals they rescued, the worse that made the spill look, and the worse it made Exxon and the government look.

The ultimate tragedy is that this oil spill was not a learning experience. Not only would more oil spills happen after this, but an oil spill far worse would happen: in the Gulf of Mexico, on April 20, 2010, an oil drilling rig called the *Deepwater Horizon* exploded, killing 11 workers, spilling 210 million gallons of oil into the Gulf, and killing far more animals than the *Exxon Valdez* spill. Meanwhile, the oil industry makes nearly 6 million dollars a day from oil just in Alaska. And the oil companies continue to have the power to pressure governments into allowing them to drill for oil and transport it through pipelines and on ships, where there is always the potential for disaster.

Most humans depend on oil to fuel our cars and airplanes, to fuel our factories, and to heat our homes. So all of us who use oil are partly responsible for oil spills such as the one described in this book—and for climate change, which is also caused by oil and is an even bigger problem than oil spills. If climate change continues at its current rate, it has the power to wipe out much of the life on this planet. Unless we learn to rely on alternate forms of energy, such as wind and solar power, our dependence on oil will make the earth uninhabitable for many life-forms, *including humans*.

FURTHER READING

Benoit, Peter. *The Exxon Valdez Oil Spill.* New York: Children's Press, 2011.

Coates, Peter. *Trans-Alaskan Pipeline Controversy.* Fairbanks, AK: University of Alaska Press, 1993.

Cohen, Stan. *The Great Alaska Pipeline*. Missoula, MT: Pictorial Histories, 1992.

Davidson, Art. *In the Wake of the "Exxon Valdez": The Devastating Impact of the Alaska Oil Spill.* New York: Random House, 1990.

Day, Angela: *Red Light to Starboard: Recalling the Exxon Valdez Disaster*. Pullman, WA: Washington State University Press, 2014.

Hartsig, Andrew, and Chris Robbins. "Exxon Valdez: 29 Years Later." *Ocean Conservancy*, March 22, 2018. https://oceanconservancy.org/blog/2018/03/22/exxon-valdez-29-years-later/.

Kernes, Susan, reporter. "Natives and the Oil Spill." *Living on Earth.* Aired week of June 11, 1999, on National Public Radio. http://www.loe.org/shows/segments.html?programID=99-P13-00024&segmentID=9.

Lee, Jane J. "On 25th Exxon Valdez Anniversary, Oil Still Clings to Beaches." *National Geographic*, March 26, 2014. https://news.nationalgeographic.com/news/2014/03/140324-exxon-valdez-oil-spill-25th-anniversary-alaska-ocean-science/.

Lynch, Michael J. "Native Americans and the Exxon Valdez Oil Spill." *Green Criminology* (blog). September 6, 2014. http://greencriminology.org/glossary/native-americans-and-the-exxon-valdez-oil-spill/.

Marsh, Laura. *Sea Otters.* Washington, DC: National Geographic Society, 2014.

Ott, Riki. *Not One Drop: Betrayal and Courage in the Wake of the Exxon Valdez Oil Spill.* White River Junction, VT: Chelsea Green Publishing, 2008.

Smith, Roland. *Sea Otter Rescue: The Aftermath of an Oil Spill.* New York: Puffin Books, 1999.

Stroymeyer, John. *Extreme Conditions: Big Oil and the Transformation of Alaska.* New York: Simon & Schuster, 1993.

Wight, Philip. "How the Alaska Pipeline Is Fueling the Push to Drill in the Arctic Refuge." *Yale Environment 360*, November 16, 2017. https://e360.yale.edu/features/trans-alaska-pipeline-is-fueling-the-push-to-drill-arctic-refuge.

Williams, Maria Sháa Tláa. *The Alaska Native Reader: History, Culture, Politics.* World Readers. Durham, NC: Duke University Press, 2009.

Wolfe, Art. *Alaska*. Seattle: Sasquatch Books, 2000.

Yergin, Daniel. *The Prize: The Epic Quest for Oil, Money, Power.* New York: Simon & Schuster, 1990.

BEACH LANE BOOKS • An imprint of Simon & Schuster Children's Publishing Division • 1230 Avenue of the Americas, New York, New York 10020

• BEACH LANE BOOKS is a trademark of Simon & Schuster, Inc. • For information about special discounts for bulk purchases, please contact Simon & Schuster Special Sales at 1-866-506-1949 or business@simonandschuster.com. • The Simon & Schuster Speakers Bureau can bring authors to your live event. For more information or to book an event, contact the Simon & Schuster Speakers Bureau at 1-866-248-3049 or visit our website at www.simonspeakers.com. • Jacket design by Rebecca Syracuse • The text for this book was set in Neutraface 2 Text. • Manufactured in China • 0120 SCP • First Edition 10 9 8 7 6 5 4 3 2 1 • Library of Congress Cataloging-in-Publication Data • Names: Winter, Jonah, 1962– author. | Winter, Jeanette, illustrator. • Title: Oil / Jonah Winter ; illustrated by Jeanette Winter. • Description: First edition. | New York : Beach Lane Books, [2019] | Audience: Ages 4-8. | Audience: Grades K-3. | Includes bibliographical references and index. • Identifiers: LCCN 2018040861 | ISBN 9781534430778 (hardcover : alk. paper) | ISBN 9781534430785 (ebook) • Subjects: LCSH: Oil spills—Environmental aspects—Juvenile literature. | Oil pollution of the sea—Juvenile literature. | Restoration ecology—Juvenile literature. • Classification: LCC TD427.P4 W578 2019 | DDC 363.738/2—dc23 LC record available at https://lccn.loc.gov/2018040861